George Nathaniel Curzon

THE ROMANES LECTURE

1907

FRONTIERS

Elibron Classics
www.elibron.com

THE ROMANES LECTURE

1907

Frontiers

BY

THE RIGHT HONOURABLE

LORD CURZON OF KEDLESTON

D.C.L., LL.D., F.R.S.

ALL SOULS COLLEGE, CHANCELLOR OF THE UNIVERSITY

DELIVERED

IN THE SHELDONIAN THEATRE, OXFORD

NOVEMBER 2, 1907

OXFORD

AT THE CLARENDON PRESS

1907

·HENRY FROWDE, M.A.

PUBLISHER TO THE UNIVERSITY OF OXFORD

LONDON, EDINBURGH

NEW YORK AND TORONTO

FRONTIERS

WHEN, at the end of December, 1905, the then Vice-Chancellor asked me to be the Romanes Lecturer in the following year, just after my return from India, I felt that the honour was one which it was impossible for me, as a devoted son of this ancient and illustrious University, to decline. But when he informed me that the entire field of Science, Literature, and Art was at my disposal for the choice of a subject, and that among my predecessors were to be found the great names of Gladstone, Huxley, and John Morley, I was more appalled at my temerity in venturing to tread in their footsteps than I was gratified at the almost illimitable range that was opened to my ambition. In these circumstances, I concluded that my best course would be to select some topic of which I had personal experience, and upon which I could, without presumption, address even this famous and learned University. I chose the subject of Frontiers. It happened that a large part of my younger days had been spent in travel upon the boundaries of the British Empire in Asia, which had always exercised upon me a peculiar fascination. A little later, at the India Office and at the Foreign Office, I had had official cognizance of a period of great anxiety, when the main sources of diplomatic preoccupation, and sometimes of international danger, had been the determination of the Frontiers of the Empire in Central Asia, in every

A 2

part of Africa, and in South America. Further, I had just returned from a continent where I had been responsible for the security and defence of a Land Frontier 5,700 miles in length, certainly the most diversified, the most important, and the most delicately poised in the world; and I had there, as Viceroy, been called upon to organize, and to conduct the proceedings of, as many as five Boundary Commissions.

I was the more tempted to undertake this task because I had never been able to discover, much less to study, its literature. It is a remarkable fact that, although Frontiers are the chief anxiety of nearly every Foreign Office in the civilized world, and are the subject of four out of every five political treaties or conventions that are now concluded, though as a branch of the science of government Frontier policy is of the first practical importance, and has a more profound effect upon the peace or warfare of nations than any other factor, political or economic, there is yet no work or treatise in any language which, so far as I know, affects to treat of the subject as a whole. Modern works on geography realize with increasing seriousness the significance of political geography; and here in this University, so responsive to the spirit of the age, where I rejoice to think that a School of Geography has recently been founded, it is not likely to escape attention. A few pages are sometimes devoted to Frontiers in compilations on International Law, and here and there a Frontier officer relates his experience before learned societies or in the pages of a magazine. But with these exceptions there is a practical void. You may ransack the catalogues of libraries, you may search the indexes of celebrated historical works, you may study the writings of scholars, and you will find the subject almost wholly ignored. Its formulae are hidden in the arcana of diplomatic chancelleries ; its documents are embedded in vast and forbidding collections of treaties ; its incidents and what I may describe as its incomparable drama are the possession of a few silent men, who may be found in the clubs of London, or Paris, or Berlin, when they are not engaged in tracing lines upon the unknown areas of the earth.

Frontiers in History.

And yet I would invite you to pause and consider what Frontiers mean, and what part they play in the life of nations. I will not for the moment go further back than a century. It was the adoption of a mistaken Frontier policy that brought the colossal ambitions of the great Napoleon with a crash to the ground. The allied armies might never have entered Paris had the Emperor not held out for an impossible Frontier for France. The majority of the most important wars of the century have been Frontier wars. Wars of religion, of alliances, of rebellion, of aggrandisement, of dynastic intrigue or ambition—wars in which the personal element was often the predominant factor—tend to be replaced by Frontier wars, i. e. wars arising out of the expansion of states and kingdoms, carried to a point, as the habitable globe shrinks, at which the interests or ambitions of one state come into sharp and irreconcilable collision with those of another.

To take the experience of the past half-century alone. The Franco-German War was a war for a Frontier, and it was the inevitable sequel of the Austro-Prussian campaign of 1866, which, by destroying the belt of independent States between Prussia and her Rhenish provinces, had brought her up to the doors of France. The campaign of 1866 was itself the direct consequence of the war of 1864 for the recovery by Germany of the Frontier Duchies of Schleswig-Holstein. The Russo-Turkish War originated in a revolt of the Frontier States, and every Greek war is waged for the recovery of a national Frontier. We were ourselves at war with Afghanistan in 1839, and again in 1878, we were on the verge of war with Russia in 1878, and again in 1885, over Frontier incidents in Asia. The most arduous struggle in which we have been engaged in India in modern times was waged with Frontier tribes. Had the Tibetans respected our Frontiers, we should never have marched three years ago to Lhasa. Think, indeed, of what the Indian Frontier Problem, as it is commonly called, has meant and means ; the controversies it has provoked, the passions it has aroused ; the reputations that have flashed or faded within its sinister shadow. Japan came to blows with China over the Frontier-state of Korea ; she found herself gripped in a life-and-death struggle with Russia because of the attempt of the latter to include Manchuria within the Frontiers of her political influence. Great Britain was on the brink of a collision with France over the Frontier incident of Fashoda ; she advanced to Khartoum not to avenge Gordon, but to defend an imperilled and to recover a lost Frontier. Only the other day the Algeciras Conference was sitting to determine the degree to which the possession of a contiguous Frontier gave France the right to exercise a predominant influence in Morocco. But perhaps a more striking illustration still is that of Great Britain and America. The two occasions on which in recent times (and there are earlier examples [1]) the relations between these two allied and fraternal peoples—conflict between whom would be a hideous crime—have been most perilously affected, have both been concerned with Frontier disputes—the Venezuelan and the Alaskan Boundary.

The most urgent work of Foreign Ministers and Ambassadors, the foundation or the outcome of every *entente cordiale*, is now the conclusion of Frontier Conventions in which sources of discord are removed by the adjustment of rival interests or ambitions at points where the territorial borders adjoin. Frontiers are indeed the razor's edge on which hang suspended the modern issues of war or peace, of life or death to nations. Nor is this surprising. Just as the protection of the home is the most vital care of the private citizen, so the integrity of her borders is the condition of existence of the State. But with the rapid growth of population and the economic need for fresh outlets, expansion has, in the case of the Great Powers, become an even more pressing necessity. As the vacant spaces of the earth are filled up, the competition for the residue is temporarily more keen. Fortunately, the process is drawing towards a natural termination. When all the voids are filled up, and every Frontier is defined, the problem will assume a different form. The older and more powerful nations will still dispute about their Frontiers with each other; they will still encroach upon

---

[1] For instance the question of the San Juan Islands in the Straits of Georgia between the mainland and Vancouver's Island, which was decided by the German Emperor in 1872 in favour of the United States; and the earlier controversies about the boundaries of Oregon and Maine.

and annex the territories of their weaker neighbours; Frontier wars will not, in the nature of things, disappear. But the scramble for new lands, or for the heritage of decaying States, will become less acute as there is less territory to be absorbed and less chance of doing it with impunity, or as the feebler units are either neutralized, or divided, or fall within the undisputed Protectorate of a stronger Power. We are at present passing through a transitional phase, of which less disturbed conditions should be the sequel, falling more and more within the ordered domain of International Law.

The illustrations which I have given, and which might easily be multiplied, will be sufficient to indicate the overwhelming influence of Frontiers in the history of the modern world. Reference to the past will tell a not substantially different tale. In our own country how much has turned upon the border conflict between England and Scotland and between England and Wales, In Ireland the ceaseless struggle between those within and those outside the Pale has left an ineffaceable mark on the history and character of the people. Half the warfare of the European continent has raged round the great Frontier barriers of the Alps and Pyrenees, the Danube and the Rhine. The Roman Empire, nowhere so like to our own as in its Frontier policy and experience,—a subject to which I shall have frequent occasion to revert,—finally broke up and perished because it could not maintain its Frontiers intact against the barbarians.

I wonder, indeed, if my hearers at all appreciate the part that Frontiers are playing in the everyday history and policy of the British Empire. Time was when England had no Frontier but the ocean. We have now by far the greatest extent of territorial Frontier of any dominion in the globe. In North America we have a Land Frontier of more than 3,000 miles with the United States. In India we have Frontiers nearly 6,000 miles long with Persia, Russia, Afghanistan, Tibet, China, Siam, and France. In Africa we have Frontiers considerably over 12,000 miles in length with France, Germany, Italy, Portugal, and the Congo State, not to mention our Frontiers with native states and tribes. These Frontiers have to be settled, demarcated, and then maintained. We commonly speak of Great Britain as the greatest sea-power, forgetting that she is also the greatest land-power in the Universe. Not much is heard of this astonishing development in Parliament; I suspect that even in our Universities it is but dimly apprehended. Nevertheless, it is the daily and hourly preoccupation of our Foreign Office, our India Office, and our Colonial Office; it is the vital concern of the greatest of our colonies and dependencies; and it provides laborious and incessant employment for the keenest intellects and the most virile energies of the Anglo-Saxon race.

My main difficulty is not how to deal with such a topic adequately, but how to deal with it in the compass of an academic lecture. As my investigations have progressed, I have seen the horizon expand before me, until it has appeared to embrace all history, the greater part of geography, and a good deal of jurisprudence. I have alternately seen my Essay swell into a volume, and have contemplated reducing that volume into a tabloid for passing consumption in this theatre. It is obviously impossible for me to treat of Frontiers in the space of an hour or of many hours; on the other hand I have found neither the leisure nor the health to write the volume which at one time I had in view. The result must be a compromise. Large portions of my subject, and indeed of my manuscript, must be ignored to-day. Before my Essay assumes a final form, I hope that it may acquire a character more in consonance with the magnitude and the unity of the theme.

I will not pause to dilate upon the obvious truisms that lie at the threshold of my subject. The influence of region upon race, and the correlative influence of race upon region, are speculations belonging to the wider subject of which Frontiers are only a part. That a country with easily recognized natural boundaries is more capable of defence and is more assured of a national existence than a country which does not possess those advantages; that a country with a sea Frontier, such as the British Isles, particularly if she also possesses sea-power, is in a stronger position than a country which only has land Frontiers and requires a powerful army to defend them ; that a mountain-girt country is the most secure of internal States—these are the common-places of political geography. More pertinent is it to say in passing that in the study of such a subject as ours, we must be very careful not to generalize too hastily as to the influence of physical agencies, either upon character or action : for the same causes are apt to produce very different results in different places or at different times. There is a passage, for instance, in an English poet which typifies what I mean. Cowper wrote in *The Task* (Book II)—

Lands intersected by a narrow forth
Abhor each other. Mountains interposed
Make enemies of nations who had else
Like kindred drops been mingled into one.

Many instances can doubtless be found in which both these propositions are true. The intervention of a narrow forth has certainly been one of the main causes of the inveterate estrangement of the English and the Irish : it has been largely responsible for the conventional hostility between England and France. But quite as many instances can be found in which the peoples on two sides of a strait or narrow sea have been on friendly terms. The generalization about mountains is equally unscientific. Nor is the inverse in either case any truer, viz. that States which are not separated from each other, either by narrow seas or by mountains, are therefore naturally friends. The fact is that in all such cases a great many causes are at work, of which geographical position or environment is but one. A safer procedure is that of deduction only from established facts. Macaulay, in his *Frederick the Great*, wrote in his pictorial manner, but with incontrovertible truth :—

Some states have been enabled by their geographical position to defend themselves with advantage against immense forces. The sea has repeatedly protected England against the fury of the whole Continent. The Venetian Government, driven from its possessions on the land, could still bid defiance to the Confederacy of Cambray from the arsenal amid the lagoons. More than one great and well appointed army, which regarded the shepherds of Switzerland as an easy prey, has perished in the passes of the Alps.

Here the philosophy of Frontiers is demonstrated by concrete facts.

Origin of Frontiers.

If we start with the dawn of history, at least in Europe, it is not difficult to trace the conditions under which the first Frontiers came into being. The existing peoples of

Europe are with rare exceptions, of whose origin we have no certain knowledge, the deposit of successive waves of human immigration from Asia. These nomads would, in each case, pursue what may be described as the groove principle of advance. Entering Europe, for the most part, through the gap between the southern end of the Urals and the Caspian, they found a continent largely overspread, particularly in its northern parts, with forests, morasses, and swamps, intersected everywhere by great rivers, and in its central and southern portions split up by mountainous masses or long projecting spurs. Between, or amid, these obstacles they would naturally follow—as they did follow—the easiest and most accessible grooves, everywhere taking advantage of natural barriers, and settling down in areas of which the limits had been provided for them. Rivers were not a natural Frontier in those primitive days. More often they were a means of access to a country than a line of division between races : indeed, both banks were not unlikely to be occupied by the same race. Only as time passed and artificial boundaries were required to supplement those of nature did rivers, though natural in origin, begin to play a part in the scheme of territorial subdivision. Mountains constituted the earliest and most obvious barriers. Then as forests were cut down and swamps were drained, the peoples pushed their way over the more level areas, until, little by little, the empty places were occupied. To what extent natural features were responsible for the earliest form of organized state may be seen from the city communities and republics of the Greek and Latin races. The limits of each were determined by its mountain barriers, and there was but little communication or power of common action between State and State.[1] It was in the more open countries that larger kingdoms and empires tended to be formed. Then, as population increased, and commerce and industry grew, as naval and military forces developed, and as larger political aggregations began to supersede the smaller units, natural boundaries were found no longer to suffice. It became necessary to supplement or to replace them by artificial Frontiers, finding their origin in the complex operations of race, language, trade, religion, and war.

Natural Frontiers.

Here, however, I must pause to define the factors and to indicate with greater precision the various classes of Frontiers with which we are called upon to deal. I have already accepted the broad distinction between Natural and Artificial Frontiers, both as generally recognized, and as scientifically the most exact. Of all Natural Frontiers the sea is the most uncompromising, the least alterable, and the most effective. The defensive attributes of the sea receive the poet's testimony in his description of England as

compassed by the *inviolate* sea ;

its exclusive qualities are celebrated in a not less familiar phrase :

the unplumbed, salt, *estranging* sea.

It is true that in one aspect the sea may be regarded as a connecting link by the artificial aid of navigation,

[1] Hence the failure of the attempt to defend Thessaly against the armies of Xerxes in 480 and the small number of troops collected to defend Thermopylae.

much in the same way as a land Frontier may be crossed by a railroad or pierced by a tunnel. The sea was such a link in the case of Greece and her city colonies, enabling her (no hostile power having command of the Mediterranean) to maintain connexion with her foreign colonies, in spite of the fissiparous influence of her physical configuration on the mainland. Similarly, the Mediterranean was a connecting link between Rome and her outlying possessions, just as it is at the present day between France and her North African colonies. Indeed the Mediterranean has never in civilized times been the southern Frontier of Europe ; the latter has in reality been supplied by the Atlas Mountains and the great Desert of Sahara. The Persian Gulf is another illustration of a sea that has shown itself to be not a barrier but a link : the Arabs who were skilled and courageous mariners from early times (witness the story of Sinbad the Sailor) having occupied both shores with their settlements. The sea may also act as a vehicle of invasion. It brought the Jutes and Saxons, the Norsemen and Danes to England, just as it had brought the Romans before them.

Nevertheless, the opposite or separating quality of the sea is undoubtedly the more striking and familiar aspect. I have already alluded to its influence upon the relations of Great Britain and France, and of Great Britain and Ireland. Only when England ceased to be a Continental power did the national spirit blossom into any fullness. The project of a Channel Tunnel between the coast of England and France has twice perished because of the invincible and legitimate repugnance of the Englishman to sacrifice his maritime Frontier for no tangible return. It was because of the

interposition of the sea that England lost America; that
the Dutch and Portuguese lost the greater part of their
Indian Empires; that Napoleon, equally with Rome,
experienced so many difficulties in Egypt; that the
Mexican adventure of France and Austria ended in
fiasco; that Spain was robbed almost in a day of her
possessions in Cuba, Porto Rico, and the Philippines.
In these circumstances the continued connexion of the
parent State with her transmarine possessions—which
might seem to be an impossible aspiration—is only
maintained in modern times by the grant of some
form of self-government with a view to attaching the
inhabitants of the colony or possession to the seat of
imperial authority.

Second in the list of Natural Frontiers may be placed
deserts, until modern times a barrier even more impassable than the sea. Asia and Africa afford the best
known instances of this phenomenon. For centuries
China has been protected by the great Gobi Desert
on her north-west border; Samarkand and Bokhara
were shielded by the sandhills of the Kara Kum; the
Indus valley was severed from the rest of India in
its southerly portions by the Sind Desert, while Beluchistan is still safeguarded by the waste that stretches
from Nushki to Seistan. Syria found a natural Frontier
on the east in the desert that bears her name. Indeed,
the whole of western Asia, that part, in fact, which was
exposed to Hellenic influences, was for centuries cut off
from India by the broad wastes of Persia and Turkestan.
Egypt, protected on the west by the impassable barrier
of the Libyan Desert, and on the east by the desert of
the Sinaitic peninsula, has retained a physical identity
almost unequalled in history. It was of her eastern

desert Frontier that the greatest captain of modern times, who had himself crossed it at the head of an army, wrote as follows in his Commentaries :—

Generals who have marched from Egypt to Syria or from Syria to Egypt have in all periods of history considered this desert the greater obstacle the larger the number of horses they took with them. The ancient historians declare that when Cambyses wished to penetrate into Egypt he made an alliance with an Arab king, who caused a canal to flow with water in the desert, which evidently means that he covered it with camels bearing water. Alexander sought to please the Jews so that they might help him in the passage of the desert. This obstacle, however, was not so great in ancient times as it is to-day, since towns and villages existed, and the industry of man contended with success against the difficulties. To-day scarcely anything remains between Salihiyeh and Gaza. An army must, therefore, cross the desert successively by forming establishments and magazines at Salihiyeh, Katieh, and El Arish. If this army starts from Syria it must first of all form a large magazine at El Arish, and then carry it forward to Katieh. But these operations are slow, and they give an enemy time to make his preparations for defence. . . . An army defending Egypt can either assemble at El Arish to oppose the investment of this place, or at Katieh to raise the siege of El Arish, or at Salihiyeh : all these alternatives offer advantages. *Of all obstacles which may cover the frontiers of empires, a desert like this is incontestably the greatest. Mountains like the Alps take second rank, and rivers the third.* If there is so much difficulty in carrying the food of an army that complete success is rarely obtained, this difficulty becomes twenty times greater when it is necessary to carry water, forage, and fuel, three things which are weighty, difficult to carry, and usually found by armies upon the ground they occupy.

Finally, Africa furnishes the crowning illustration of the Great Sahara, which for centuries not only cut off the Mediterranean belt from the rest of Africa, but cut off the entire remainder of Africa from the civilized world. It was not till the voyages of Prince Henry the Navigator that this long period of isolation came to an end.

In quite recent days, however, deserts as Frontiers have lost the greater part both of their terror and their strength. Like the loftiest mountains, like the stormiest oceans, they have yielded to the all-conquering influence of steam. When Skobeleff advanced to the extirpation of the Tekke Turkomans at Geok Tepe in 1881, he laid a light railway behind him. As the English troops moved up the Nile to the recovery of Khartoum in 1898, the steel rails kept pace with their advance. It was to the desert line, and not to the actual collision at Omdurman, that the Khalifa owed his destruction. The Turkish railways to Syria, if protracted, would soon deprive Egypt of the security on her eastern border which Napoleon described. Given a level desert with a sound foundation, a railroad becomes the easiest of constructions, and built by competent engineers, can be pushed forward in war time at the rate of three miles a day.[1] Similarly it is the aid of railways that has enabled the United States of America to make so light of the great deserts that stretch eastwards from the Rocky Mountains. There are few desert Frontiers in the world that now remain intact. If occasion arose, the doom of the survivors would probably be sealed. It is a question, not of mechanics, but of water and fuel. On the other hand, a ghost of the idea may be said to survive in the still existing preference of uninhabited or thinly peopled tracts as Frontiers over areas of crowded occupation.

I turn to the consideration of the third type of Natural Frontier, namely, mountains. We have already seen that mountains were the earliest of the barriers

[1] In the Soudan campaign of 1898, 5,300 yards of railway were surveyed, embanked, and laid in a single day.

accepted by wandering man. *Prima facie*, also, they are the most durable and the most imposing. They are liable to little change (except such as may be indirectly effected by human agency in the shape of roads, railroads, and tunnels), and they are capable of instant and easy recognition. Such has been the position, and the decisive influence of the great mountain barriers of the world—of the Hindu Kush and Himalayas in Asia,[1] of the immense and serried ridges that separate Burma from China, of the Caucasus between Asia and Europe, of the Taurus in Asia Minor, of the Alps and Pyrenees (and in a lesser degree the Balkans and Carpathians) in Europe. From the military point of view the labour in crossing a mountain range is commonly great, particularly for armies, and is exposed to many dangers. What it was in ancient times, a century ago, and in more modern days, may be compared in the crossing of the Alps by Hannibal, and again by Napoleon, and in the experience of the Russians in the Balkans. We have yet to see a campaign in which the appliances of modern science in respect of railroads, tunnels, and telegraphs, are seized by one party so as practically to annihilate the mountain barrier upon which the other relies for protection, although the Germans seized and utilized the French tunnel through the Vosges at the commencement of the Franco-German War, and the Boers possessed themselves of the Drakensberg range and the Laing's Nek tunnel in order to facilitate their descent upon Natal.

On the other hand the theoretical superiority of a

---

[1] Backed as they are by the huge and lofty plateau of Tibet, the Himalayas are beyond doubt the most formidable natural Frontier in the world.

mountain Frontier may be qualified by a number of considerations arising from its physical structure. Of course a range or ridge with a sharply defined crest is the best of all. But sometimes the mountain-barrier may be, not a ridge or even a range, but a tumbled mass of peaks and gorges, covering a zone many miles in width (for instance, the breadth of the Himalayas north of Kashmir is little short of 200 miles), and within this area the inhabitants may be independent or hostile. Such has been the case with a large portion of the Pathan Frontier of India, where the physical conformation of the border lends an immense advantage to the holders of the mountains against the occupants of the plains. The desire to counteract this advantage and to transfer it to the Cis-border Power has led to the pursuit of what is known as the Scientific Frontier, i. e. a Frontier which unites natural and strategical strength, and by placing both the entrance and the exit of the passes in the hands of the defending Power, compels the enemy to conquer the approach before he can use the passage. It is this policy that has carried the Indian outposts to Lundi Khana, to Quetta, and to Chaman, all of them beyond the passes, whose outer extremities they guard.

In every mountain border, where the entire mountainous belt does not fall under the control of a single Power, the crest or water-divide is the best and fairest line of division ; for it is not exposed to physical change, it is always capable of identification, and no instruments are required to fix it. But it is not without its possible drawbacks, of which the most familiar is the well-known geographical fact that in the greatest mountain systems of the world, for instance, the Himalayas and the Andes, the water-divide is not identical with the highest crest,

but is beyond it and at a lower elevation. Another
parent of much controversy is the ambiguous phrase
'foot of the hills', the different interpretations of which
in fixing the Indian Frontier almost produced a rup-
ture between the Indian Government and the late Amir
of Afghanistan.

We now come to the important category of Rivers.
As the creation of nature, in contradistinction to the
creation of man, no Frontiers are more natural. But in
another sense, namely, that which is in accord with the
natural habits of man, rivers are not natural divisions,
because people of the same race are apt to reside
on both banks. Thus the Germans are found living
on both banks of the Upper Danube, and the Slavs
on both banks of the Lower Danube; the Turks or
Tartars live on both banks of the Oxus, though to
the north they are in Russian (i. e. Bokharan) territory
and to the south under Afghanistan; the Indus was
not a natural frontier to the Punjab, because Indian
peoples, as distinct from Pathans or border men,
inhabit the further as well as the nearer bank of the
river. So many of the peoples of Laos lived astride
the Mekong that the French soon found it to be an
impracticable Frontier. The long contest between
America, France, and Great Britain over the Mississippi
valley, arising out of Napoleon's policy of shutting in
the United States behind the Alleghanies, and using
the Indians as a barrier between the mountains and
the river, abundantly illustrated the futility of a river
as a permanent boundary. As soon as the Americans
broke through the !Alleghanies with railroads there
was no stopping them by the Mississippi or any other
natural Frontier until they reached the sea. In fact the

teaching of history is that rivers connect rather than separate. Strategical reasons have almost invariably been responsible for their conversion into Frontiers. As States developed and considerable armies were required for their defence, the military value of rivers, in delaying an enemy, and in concentrating defensive action at certain bridges, or fords, or posts, became apparent, and in the demarcation of larger kingdoms and States, they provided a convenient line of division, everywhere recognizable, and easily capable of defence. It was for this reason that Augustus selected rivers —the Rhine and the Danube—as the Frontiers of the Roman Empire, though strategical considerations soon tempted the Romans beyond, as the English have been tempted across the Indus, and the French by other causes across the Mekong.

Accordingly the advantages and disadvantages of rivers as Frontiers may be thus stated. The position of the river is unmistakable, no survey is required to identify or describe it, and the crossing-places frequently admit of fortification. Rivers are lines of division as a rule very familiar to both parties, and are easily transferred to a treaty or traced on a map. On the other hand, they may be attended by serious drawbacks, confronting diplomatists and jurists with intricate problems. Rivers are liable to shift their courses, particularly in tropical countries. The vagaries of the Helmund in Seistan, where it is the boundary between Persia and Afghanistan, have led to two Boundary Commissions in thirty years. The precise channel which contains the Frontier line, the division of islands, very likely new accretions, in the river-bed, the determination of drinking-rights or of water-rights in cases where cultivation is only effected by means of irrigation from the Frontier river, the exact identity of the source of a river, if this be mentioned in a Treaty or Convention, or of its main affluent, or, in a deltaic region, of its mouth, the provision required for navigation, police, and fiscal control—all of these suggest possible difficulties in the acceptance of a river boundary, particularly in new or tropical countries, which cannot be ignored. In ancient and civilized States the procedure to be followed in many of these cases is regulated by international agreement or by the Law of Nations. The general principles regarding the navigation of rivers traversing different States were indeed embodied in Articles 108-116 of the Final Act of the Congress of Vienna,[1] and have been applied by subsequent Agreements to some of the principal rivers of Europe, Africa, and America.

The last Natural Frontier to which I need here refer is the wellnigh obsolete barrier created by forests and marshes and swamps. The various Saxon kingdoms of England were, for the most part, thus severed from each other. When Caesar first landed in Britain, the head quarters of Cassivelaunus, the British leader, were placed at Verulamium, near St. Albans, which was surrounded by forests and swamps. Arminius gained his famous victory over the Roman legions by entangling them in the forests and morasses of Westphalia, and there is no more poignant picture in history than the description of Tacitus of those mournful scenes.[2] Venice was many times saved from absorption both by foreign invaders, such as Goths and Huns, and by jealous neighbours by her cincture of lagoons.

[1] Vide Sir E. Hertslet's *Map of Europe by Treaty*, vol. i. p. 269.
[2] *Ann.* i. 60-71.

These, however, as cultivation, settlement, and drainage have advanced, are disappearing types of Frontier, of which no more need now be said.

Artificial Frontiers.

From Natural Frontiers I pass to the category of Artificial Frontiers, by which are meant those boundary lines which, not being dependent upon natural features of the earth's surface for their selection, have been artificially or arbitrarily created by man. These may be classified as ancient and modern, the distinction between them—which is one of method only and not of principle—roughly reflecting the difference between the requirements of primitive and of civilized peoples. Primitive society, where not assisted by natural features in the determination of its limits of occupation or conquest, but being nevertheless desirous to protect its boundaries from external aggression, commonly either erected a barrier or created a gap. Under one or other of these headings will be found to fall all the Artificial Frontiers of the ancient and mediaeval world.

The commonest type of the barrier-frontier was a palisade or mound or rampart or wall ; elsewhere use might be made of an existing road or canal or ditch. Of the latter class a familiar illustration is the great Roman road of Watling Street, in this country, which, by the Treaty of Wedmore (878 A. D.), was made the boundary between the English territories of Alfred and the Danes. An early English example of the other type was Offa's Dyke, the huge earthwork constructed by the Mercian king of that name (about 780 A. D.) from the mouth of the Wye to the mouth of the Dee as a Frontier against the Welsh. Palisades have been found as far apart as the borders of China and Manchuria and of Manchuria and Korea, and the outskirts of the Roman dominions beyond the Danube and the Rhine. Spartianus, in his Life of Hadrian,[1] describes the palisade erected by that Emperor in the trans-Danubian section of the Roman Frontier, and the researches of the German explorers, who in recent years have laid bare the traces of that remarkable barricade from end to end, have revealed the existence of split oak trunks, nine feet high, driven into a deep ditch, and held together by stout transverse beams.

The palisade or rampart or wall of ancient history was, however, the commonest illustration of a type of Frontier that was concerned less with delimitation than with defence. When the Chinese built the Great Wall of China, when Hadrian and Antoninus Pius and Severus raised the double line of fortification between the Firths of Clyde and Forth, and between the Solway and the mouth of the Tyne, when the Flavian Emperors built the Pfahlgraben and other ramparts or walls between the Rhine and the Danube, when the successors of Alexander raised a similar barrier in the country to the east of the Caspian—one and all were not thinking so much of rounding off the territories of conquests of the Empire as they were of protecting its Frontiers in the best manner against the terrible and ever-swelling menace of the barbarians. Consequently the wall or barrier was sometimes erected upon the administrative Frontier, and sometimes far in advance of it. Though Hadrian's wall was for centuries the effective Frontier of the Roman dominion in Britain, the Romans yet to some

[1] *Vit. Hadriani*, 12.

extent occupied the ground between it and the second or northern wall, and even beyond the latter. Similarly the Roman *limes* in what is now Germany and Austria advanced or receded, not so much as indicating a fluctuating movement of the real boundaries of the Empire, as following the best line of military defence that was suggested by the exigencies of the time. Trajan's conquest of Dacia was of course a positive, though only a temporary, extension of the Empire.

Rudimentary in conception though these structural barriers may be thought to have been, they were effective in the age and against the foes for whom they were devised. There can be no greater mistake than to ridicule them as monuments of misdirected effort or of human vanity. The Great Wall of China, commenced before the Christian Era and continued at intervals for 1,700 years, was a genuine palladium to the heart of the Chinese Empire. Though occasionally circumvented and more than once pierced by the nomad hordes, for centuries it held back the Mongolian Tartars from Peking, acting as a fiscal barrier for the prevention of smuggling and the levying of dues, as a police barrier for the examination of passports and the arrest of criminals or suspects, and as a military barrier against hostile invasions or raids. It was even more a line of trespass than a Frontier. Much the same might be said of the Roman walls, whether directed against the Picts and Scots, or against the Marcomanni and Teutonic tribes. Guarded by fortified posts or forts at intervals, with watch-towers between, garrisoned in the early days of the Roman Empire by veterans of the army, in later times by native auxiliaries, with the great legionary camps in the rear, they kept the front of the

Empire until the ever-mounting crest of the barbarian torrent burst through defences, which there were no longer the men or the military spirit to defend. Walls and ramparts have now passed away as Frontiers of dominions, just as they are becoming obsolete as defences of cities. Occasionally in some remote corner to which the tides of human movement have not penetrated, their survivals are found. The Tibetans thought that they could bar their mountain plateaux to the Indian army by a stone wall built across a valley. A more practical and modern analogy has been traced in the Customs Hedge or Frontier made of thorny bushes and trees, which until a few years ago was stretched for 2,500 miles round the territories of British India to keep out contraband salt from the Native States.[1]

But a commoner and more widely diffused type of ancient Frontier was that of the intermediary or Neutral Zone. This may be described as a Frontier of separation in place of contact, a line whose distinguishing feature is that it possesses breadth as well as length. Sometimes it was a razed or depopulated or devastated tract of country; at others a debatable strip between the territories of rival powers : or, again, a border territory subject to and defended by one party, though exposed to the ravages of the other. Between Korea and China there existed, till beyond the middle of the last century, a broad uncultivated and uninhabited tract over 5,000 square

---

[1] Such devices are not unknown in Europe. A few years ago the smuggling of tobacco, sugar, and salt across the Swiss Frontier into Italy, where heavy duties are imposed, was so incessant that the Italian Government fenced off a large portion of the Frontier with wire netting, the gates of which were fitted with alarm-bells, and planted along it a cordon of Customs House officers armed with rifles.

miles in extent in which neither people were permitted to settle under penalty of death (though latterly it began to be encroached upon by colonists from both sides), and which became an Alsatia for roving banditti. A more innocent reproduction of the same idea may still be seen in the Neutral Ground between the British and Spanish Lines at Gibraltar. Travellers have reported the existence of the same device for keeping apart the lands of tribal communities in Central Africa, in the interior regions of the Soudan, the Congo, and the Niger.

In mediaeval times we see a more developed form of the same expedient in the Marks or Marches—a part of the settled policy of Charlemagne and Otto, and generally of the Frankish and German kings. From these Marks, intended to safeguard the Frontiers of the Empire from Slavonic or alien contact, and ruled by Markgrafs or Margraves, sprang nearly all the kingdoms and States which afterwards obtained an independent national existence, until they became either the seats of empires themselves, as in the case of the Mark of Brandenburg, or autonomous members of the German Federation. The same word (already familiar in the Marcomanni or Marchmen of the days of Antoninus), and in slightly different forms the same practice, reappear in the West Saxon kingdom of Mercia, or the March-land, in the Marches between England and Wales, for five centuries the scene of bloody conflicts between the Marcher Lords or delegates of the English kings, and the Welsh inhabitants; in the title Marquis, springing from that office; in the Wardens of the Marches, three on the English and three on the Scottish border, who watched each other from both sides of the Tweed and the Cheviots, and interwove a woof of chivalry and high romance into a warp of merciless rapine and savage deeds; and, finally, in the title of the Merse given to the Lowland counties on the Scottish side of the border. I would that I had the time to say something here about that typical illustration of Frontier character and Frontier civilization, whose history is written in the battered castles and peels of the Border, and enshrined in a literature of inimitable charm.

More pertinent is it to notice that from this ancient and mediaeval conception of a neutral strip or belt of severance has sprung the modern idea of a deliberately neutralized territory, or state, or zone. The object in both cases is the same, viz. to keep apart two Powers whose contact might provoke collision : but the *modus operandi* is different. Where primitive communities began by creating a desert in order to prevent occupation, and then provided for occupation by authorities and forces specially deputed for the purpose, modern States construct their buffer by diplomatic conventions, and seek the accommodating sanction of International Law. At one end of these devices we are but little removed from primitive practice. I allude to the arbitrary and often anomalous creation by modern powers of small neutral zones, ostensibly with a view of avoiding contact, quite as frequently in order to evade some diplomatic difficulty or to furnish material for future claims. Of such a character was the 25-kilomètre strip on the right bank of the Mekong created by the Franco-Siamese Treaty of 1893, nominally owned but not policed by Siam, containing both authorities and inhabitants whose connexion lay with the opposite or French bank. Such a diplomatic fiction could only be a temporary expedient preluding a more effective solution. Similar in character and result were the neutral zone established by Great Britain and Germany in the Hinterland of the Gold Coast in 1888, and the petty buffer State which Lord Rosebery sought to erect in 1893–5 between the borders of India and of France on the Upper Mekong. The abortive Agreements of 1894 with King Leopold and the Congo State were similarly intended to set up a buffer between the Central African territories of Great Britain and her European rivals. A yet further illustration would be the Cis-Sutlej strip of territory, into which, though it belonged to the Sikh rulers, they were not permitted by treaty to enter with armed force, and their violation of which led to the Sikh wars. Of these artificial expedients it may be laid down that they have no durability, unless they are based upon some intelligible principle of construction or defined by a defensible line, and are administered by an authority capable of preserving order.[1]

The history of British, and subsequently of American, expansion in America affords a significant illustration of the futility of an artificial buffer. When the British had conquered Canada and all the territories east of the Mississippi, George III issued a Proclamation (Oct., 1763) forbidding settlements beyond the sources of the rivers flowing into the Atlantic (this was known as the 'fall line'), reserving the lands beyond the

---

[1] Such a buffer zone still exists in the shape of Wakhan, a narrow strip of territory only a few miles in width containing the upper waters of the Oxus as far as its source, which was interposed by the Convention of 1895 between the Pamir Frontier of Russia and the Hindu Kush Frontier of the Indian Empire. It can only last so long as the Amir of Afghanistan, to whom it was handed over, with a special subsidy from Great Britain, fulfils his undertaking to maintain order.

Alleghanies for the Indians, and forbidding settlers to enter them. At the same time an effort was made to construct a buffer along the Indian border by the purchase of Indian lands and the settlement of European colonists upon them. But the Alleghanies were crossed almost before the ink on the Proclamation was dry; and, these once passed, no other physical barrier intervened until the Great Plains and the Rocky Mountains, which retarded, but did not finally impede, the American advance to the Pacific.[1] The Americans inherited the British policy, and as they pushed forward kept steadily thrusting the Indian Frontier backwards by a series of removals or deportations, the object being in each case to separate them from contact with the white man. But the progress of the latter was so rapid that these artificial Frontiers were continually being caught up and overlapped, the Indian territories finding themselves enveloped in the advancing tide. This led to the 'reservation' system, which continues to this day, and under which the national existence of the Indians is only, and that with difficulty, preserved by the creation of 'enclaves' with arbitrary Frontiers.

Much more is to be said for the buffer State as commonly understood, i.e. the country possessing a

[1] A curious analogy to the experience of Great Britain in the East Indies may be observed in the nomenclature of the Frontier. The old American North West or Frontier of a century ago (Ohio, Indiana, Illinois, Michigan, and Wisconsin) is now the middle region and heart of the United States. Just so in British India, what were the North-West Provinces seventy years ago have been swallowed up in the interior, and the title has passed, with the geographical fact which it represents, to the new North-West Frontier Province, beyond the Indus, which I was responsible for creating in 1901.

national existence of its own, which is fortified by the territorial and political guarantee, either of the two Powers between whose dominions it lies and by whom it would otherwise inevitably be crushed, or of a number of great Powers interested in the preservation of the *status quo*. The valley of the Menam, which is the central portion of Siam, has been thus guaranteed by Great Britain and France; Abyssinia has been guaranteed by these two Powers and Italy: in the Agreement just concluded between Great Britain and Russia about Central Asia the integrity and independence of Persia are once more guaranteed by the two great contracting parties—thereby constituting that country a true buffer State between their respective dominions: but a new provision is introduced in which, while rival spheres of interest, Russian and British, are created on the north and east respectively, a zone is left between them of equal opportunity to both Powers. This is an arrangement wanting both in expediency and permanence, the more so as the so-called neutral zone is carved exclusively out of the regions in which British interests have hitherto been and ought to remain supreme. The same Agreement contains a further novelty in international diplomacy, in the shape of a neutralizing pledge about Tibet made by two Powers, one of which is contiguous while the other has no territorial contact whatever with that country. Tibet is not a buffer State between Great Britain and Russia; the sequel of the recent expedition has merely been to make it again what it had latterly ceased to be, namely, a Mark or Frontier Protectorate of the Chinese Empire. There is a second type of buffer State lying between two great Powers in which the predominant political influence is acknowledged to belong to one of the two and not to the other. Korea, which has passed under the unchallenged influence of Japan, is such a buffer State between Japan and China. Afghanistan is in the same position between Great Britain and Russia. Here we have a close analogy to the Mark system of the Frankish Emperors and to the practice of the Roman Empire, which sought to protect its Frontiers by a fringe of dependent kingdoms, or client-states.

In all these cases the buffer State is an expedient more or less artificial, according to the degree of stability which its government and institutions may enjoy, constructed in order to keep apart the Frontiers of converging Powers. That such an experiment must necessarily fail is certainly not the teaching of European history, where it is the marchlands of Empire, Castile, France, Prussia, that have often fought their way to greatness and fame. A quite abnormal type of a buffer State existed in Europe, with slight breaks, for over 1,100 years, in the shape of the Papal Dominions. Its boundaries were perpetually shifting, and they made no pretence to represent natural or geographical lines of division. But they approximately severed the continental and peninsular portions of Italy, those parts which have always looked to the West, and those parts which have constantly been exposed to Hellenic or Eastern influences.

In Europe, however, buffer States have not as a general rule been situated between the territories of Powers possessing superior force and a higher civilization. In Asia, where the experiment is now more commonly tried, and where the conditions are less favourable, the degree of vitality to be expected is less. There the buffer conditions are apt to foster intrigue outside, apathy and often anarchy within ; and either partition follows or the stronger and least scrupulous of the bordering Powers absorbs the whole. It is unlikely that Korea will permanently retain even her present figment of independence. The future of Siam, Persia, and Afghanistan constitutes one of those problems on which speculation on an occasion like the present would be at once improper and unwise.

Lastly, there are the States, situated entirely in Europe, which are protected by an International Guarantee. These are, Switzerland neutralized by eight Powers in 1815, Belgium neutralized by five Powers in 1831, and Luxemburg neutralized by five Powers in 1867, the object in each case having been to create a buffer State between Germany and France.[1] Neutralization does not absolutely protect, and has in practice not protected, these countries from violation: but it renders aggression less likely by making it an international issue. The desire to extend a Frontier at the expense of a neutralized State can, therefore, only be gratified at a rather expensive price. Whether Holland, and the Scandinavian kingdoms, can permanently retain their independence without the safeguard of some such form of guarantee is problematical. The former has to some extent protected itself by constituting the Hague the seat of an International Tribunal, which claims to be interested in the preservation of

---

[1] Greece, which was guaranteed by three Powers, viz. Great Britain, France, and Russia in 1830, is in a different position. It is not a buffer State, and the political and territorial guarantee of the three Powers does not prevent hostilities with any other Power.

peace: the latter are temporarily safeguarded by dynastic alliances. But in both groups lie critical Frontier issues of the future.

I now proceed to the examination of the commoner forms of Artificial Frontiers in use among modern States. These are three in number: (1) what may be described as the pure astronomical Frontier, following a parallel of latitude or a meridian of longitude; (2) a mathematical line connecting two points, the astronomical coordinates of which are specified; and (3) a Frontier defined by reference to some existing and, as a rule, artificial feature or condition. Their common characteristic is that they are, as a rule, adopted for purposes of political convenience, that they are indifferent to physical or ethnological features, and that they are applied in new countries where the rights of communities or tribes have not been stereotyped, and where it is possible to deal in a rough and ready manner with unexplored and often uninhabited tracks. They are rarely found in Europe, or even in Asia, where either long settlement or conflict has, as a rule, resulted in boundaries of another type.

(1) The best known illustration of the astronomical line is the Frontier between Canada and the United States, which from the Lake of the Woods follows the 49th parallel of latitude to the Pacific coast, a distance of 1,800 miles. This line well illustrates both the strength and the weakness of the system. As a conventional line through unknown territories it has answered its purpose. But its demarcation on the spot was so laborious and protracted that, fifty years after the conclusion of the Treaty which created it, the joint surveyors were still at work, clearing a strip 100 yards wide

through the primaeval forest, and ornamenting it with iron pillars and cairns, at a cost to both countries which was enormous. Similar lines have been employed to define the boundaries of Canada and Alaska, to separate many of the Australian Colonies from each other, to determine European Spheres of Influence or Protectorates in Africa, and, quite recently, to define the Russian and Japanese shares of the island of Saghalin. Such lines are very tempting to diplomatists, who in the happy irresponsibility of their office-chairs think nothing of intersecting rivers, lakes, and mountains, or of severing communities and tribes. But even in the most favourable circumstances they require an arduous triangulation on the spot, and until surveyed, located, and marked out, have no local or topographical value.

(2) The straight line from point to point is also a method very popular in America, where it has been employed in laying down the internal Frontiers of States, and is in keeping with the mathematical pre-.cision commonly applied to the laying out of cities and streets. Like the Frontiers of latitude or longitude this type of boundary is a useful and sometimes an indispensable expedient; but it possesses no elasticity, and it is apt to produce absurd and irrational results. It is said in America that many men reside in one State and do their business in another, and there is no reason why so artificial a device should not have even more inconvenient consequences. The internal administrative or county boundaries of Great Britain have been constructed on the opposite principle, and represent a combination of historical, geographical, and occasionally ethnological conditions.

(3) Frontiers by reference, i. e. Frontiers defined as

running in a specified direction or for a certain distance, or as the arc of a circle, or as the tangent to a circle, are very familiar features in African treaties, where use has to be made of visible features or landmarks. But they are a fruitful source of error or misunderstanding, both in terminology and topography. For instance, the Alaskan dispute between Canada and the United States turned upon the meaning of the ambiguous words 'a line parallel to the windings of the coast which shall never exceed the distance of ten marine leagues therefrom'. What was the coast referred to, and what was the practicability or meaning of a line that scaled inaccessible peaks and was lost amid ice and eternal snow ?

I have now passed in review the various forms of Frontiers either furnished by nature or created by man, and have endeavoured to indicate their degrees of strength or weakness. In practice the tendency of mankind has been to ignore or override nature, and in the case of older States to adopt racial or linguistic or purely political lines of division, in the case of the partition of new countries to adopt the temporary or conditional expedients which have been described. In North America few of the internal boundaries correspond to any natural feature. In South America where, owing to the configuration and history of the country, natural boundaries are commoner, there is scarcely an undisputed Frontier. In Europe, apart from certain ranges of mountains, (but few rivers), which being genuine barriers have exercised a permanent influence upon the formation of States and the distribution of men, the boundaries of the majority of States are purely political, and find their origin in the events of history ; although geographical conditions, such as the eligibility of elevated, or sterile, or sparsely peopled tracts for Frontier purposes, have not been without influence in their selection. Of political or historical Frontiers I may mention as illustrations the Frontiers of Spain and Portugal, of Germany and Holland, of Germany and Austria, of Russia and Germany, of France and Germany (except the Vosges), of France and Belgium, of Belgium and Holland, of Turkey and Greece. Here and there exists a State like Switzerland, which accords with no one principle of national distribution, or Belgium, which infringes them all. Both are the creations of expediency or of fear. In Asiatic Frontier delineations tribal boundaries, except where overridden by political considerations, are apt to be observed.

Modern Expedients.

In the last quarter of a century, largely owing to the international scramble for the ownerless or undefended territories of Africa and Asia, fresh developments have occurred in the expansion of Frontiers, of which notice must here be taken. The result may be much the same as in the ruder days of Alexander or Trajan or Justinian, when there was no sanction beyond that of might; modern usage, however, evolves with convenient rapidity a Law of Nations that is held to justify these more recent manifestations of the centripetal tendency of international borders. All the expedients to which I am about to refer are variations in differing stages of the doctrine of Protectorates which has existed from the remotest days of Empire.

A Protectorate is a plan adopted for extending the political or strategical as distinct from the administrative

Frontier of a country over regions which the protecting Power is, for whatever reason, unable or unwilling to seize and hold itself, and, while falling short of the full rights of property or sovereignty, it carries with it a considerable degree of control over the policy and international relations of the protected State. It involves the obligation to defend the latter from external attack, and to secure the proper treatment of foreign subjects and property inside it. To what extent it justifies interference in the internal administration of the State is a question admitting of no law.

The Roman Empire is the classical illustration of this policy, though in a somewhat inchoate form, in the ancient world. In the Western Empire Protectorates, strictly so called, were not required because the enemy with whom contact was to be avoided was the barbarian, formidable not from his organization, but from his numbers ; and against this danger purely military barriers, whether in Britain, Gaul, Germany, or Africa, required to be employed. But in the East, where the ambitions of Rome for the first time encountered a rival and civilized Power of almost equal strength with itself, namely, the Parthian or Persian Kingdom, the perils of actual contact were for long delayed by a barrier of protected States, the majority under the political suzerainty of Rome, some of them oscillating from one allegiance to another, according to the degree of pressure applied ; the most important of all, by reason of its physical features and geographical position, namely, Armenia, having a career which in its stormy vicissitudes has recalled to many writers the chequered and fateful experience, between the rival Powers of Great Britain and Russia, of the buffer kingdom of Afghanistan. The Ottoman Empire in Europe in the sixteenth and seventeenth centuries, when its fortunes began to decline from their zenith, adopted an analogous policy of client-states in the Christian territories of Transylvania, Moldavia, Wallachia, and Ragusa.

It has been by a policy of Protectorates that the Indian Empire has for more than a century pursued, and is still pursuing, its as yet unexhausted advance. First it surrounded its acquisitions with a belt of Native States with whom alliances were concluded and treaties made. The enemy to be feared a century ago was the Maratha host, and against this danger the Rajput States and Oude were maintained as a buffer. On the North-west Frontier, Sind and the Punjab, then under independent rulers, warded off contact or collision with Beluchistan and Afghanistan, while the Sutlej States warded off contact with the Punjab. Gradually, one after another, these barriers disappeared as the forward movement began : some were annexed, others were engulfed in the advancing tide, remaining embedded like stumps of trees in an avalanche, or left with their heads above water like islands in a flood. When the annexation of the Punjab had brought the British power to the Indus, and of Sind, to the confines of Beluchistan ; when the sale of Kashmir to a protected chief carried the strategical Frontier into the heart of the Himalayas ; when the successive absorption of different portions of Burma opened the way to Mandalay, a new Frontier problem faced the Indian Government, and a new ring of Protectorates was formed. The culminating point of this policy on the western side was the signature of the Durand Agreement at

Kabul in 1893, by which a line was drawn between the tribes under British and those under Afghan influence for the entire distance from Chitral to Seistan, and the Indian Empire acquired what, as long as Afghanistan retains an independent existence, is likely to remain its Frontier of active responsibility. Over many of these tribes we exercise no jurisdiction, and only the minimum of control; into the territories of some we have so far not even penetrated ; but they are on the British side of the dividing line, and cannot be tampered with by any external Power. My own policy in India was to respect the internal independence of these tribes, and to find in their self-interest and employment as Frontier Militia a guarantee both for the security of our inner or administrative border, and also for the tranquillity of the border zone itself. Further to the east and north the chain of Protectorates is continued in Nepal, Sikkim, and Bhutan : on the extreme north-east the annexation of Upper Burma has brought to us the heritage of a fringe of protected States known as the Upper Shan States. At both extremities of the line the Indian Empire, now vaster and more populous than has ever before acknowledged the sway of an Asiatic sovereign, is only separated from the spheres of two other great European Powers, Russia and France—the former by the buffer States of Persia and Afghanistan and the buffer strip of Wakhan ; the latter by the buffer State of Siam, and the buffer Protectorates of the Shan States. The anxiety of the three Powers still to keep their Frontiers apart, in spite of national *rapprochements* or diplomatic *ententes*, is testified by the scrupulous care with which the integrity of the still intervening States is assured, and, in the case of this country, by the enormous sums

that have been spent by us in fortifying the independence of Afghanistan. The result in the case of the
Indian Empire is probably without precedent, for it gives
to Great Britain not a single or double but a threefold
Frontier, (1) the administrative border of British India,
(2) the Durand Line, or Frontier of active protection,
(3) the Afghan border, which is the outer or advanced
strategical Frontier.

It may be observed that the policy of Protectorates
which I have described is by no means peculiar even in
modern times to Great Britain, although Great Britain,
owing to the huge and vulnerable bulk of her Empire,
supplies the most impressive modern illustration. The
policy has been equally adopted by Russia and by
France. The Russian scheme of Protectorates includes
Khiva and Bokhara : it aims at Mongolia : it broke down
from the attempt to incorporate Manchuria and Korea.
The French Protectorates in Africa embrace Tunis and
would fain embrace Morocco ; in Asia they veil with the
thinnest of disguises the practical absorption of Cambodia
and Annam. Protectorates are also a familiar expedient
in the partition of Africa by European Powers, although
the phrase more commonly applied in those regions is
the less precise definition of a 'sphere of influence'.
With what varied objects these different Protectorates
have been established, sometimes political, sometimes
commercial, sometimes strategic, sometimes a combination of all these, I have not time here to deal. But three
curious and exceptional cases may be mentioned : that
of the British Somaliland Protectorate, acquired in
order to safeguard the food-supply of Aden (just as the
Roman Protectorate was extended over Egypt, in order
to ensure the corn-supply of Rome), and the British

Protectorate of the petty Arab chiefships on the southern shore of the Persian Gulf, established in order to prevent slave-raiding on the adjoining seas. The third case is the anomalous and unprecedented form of Protectorate declared by the United States of America in the extreme assertion of the Monroe doctrine over the Latin States of Central and Southern America. This Protectorate appears to involve a territorial guarantee of the States in question against any European Power; but what measure of internal control or interference it may be held to justify, no man can say.

Protectorates shade away by imperceptible degrees into the diplomatic concept now popularly known as Spheres of Influence. When first this phrase was employed in the language of diplomacy I do not know, but I doubt if a more momentous early use of it can be traced than that in the assurance first given by Count Gortchakoff to Lord Clarendon in 1869, and often since repeated, that Afghanistan lay 'completely outside the sphere within which Russia might be called upon to exercise her influence'. Since those days Spheres of Influence have become, notably in Africa, though scarcely less in Asia, one of the recognized means of extending a Frontier or of pegging out a potential claim. A Sphere of Influence is a less developed form than a Protectorate, but it is more developed than a Sphere of Interest. It implies a stage at which no exterior Power but one may assert itself in the territory so described, but in which the degree of responsibility assumed by the latter may vary greatly with the needs or temptations of the case. The native Government is as a rule left undisturbed; indeed its unabated sovereignty is sometimes specifically re-affirmed; but commercial exploitation and political influence are regarded as the peculiar right of the interested Power. No body of rules can, however, be laid down : for it is obvious that a Sphere of Influence in a still independent kingdom like Persia, must be a very different thing from a Sphere of Influence among the semi-barbarous tribes of the Bahr-el-Ghazel or the Niger.

Some of the most anxious moments of modern history have arisen from the vague and grandiose interpretation given to this claim by modern Powers.[1] Sometimes the advance has been so rapid that even the inner Frontiers of the Sphere of Influence are unknown,[2] at others the claim itself is so vast that it can be supported neither by reasoning nor by force. At the same time, in an epoch when the rate of advance has so frequently been in inverse proportion to the means for effective occupation or the capacity of military defence, it is probable that on the whole a pacific influence has been exercised by the diplomatic recognition of these somewhat anomalous types, which usually present one very remarkable and highly characteristic feature—that they are constructed by European statesmen with the minimum of reference or deference to the parties *prima facie* most interested, namely, the occupants of the sphere itself.

The theory of Hinterland is another modern application of the doctrine of Spheres of Influence, resting the

---

[1] In this connexion may be recalled the arrest of Colonel Young-husband by a Russian detachment under Colonel Yonoff in 1891 at Bozai Gumbaz on the Upper Oxus, on the ground that he was trespassing upon the then unknown and undefined Russian Sphere of Influence in the Pamirs.

[2] In 1893 an unfortunate collision, in which European lives were lost, occurred at Waima in the Hinterland of Sierra Leone between English and French forces, as to which neither side could state with certainty in whose sphere the scene of the disaster lay.

case for an advance of Frontier on the ground of territorial continuity. In one sense the doctrine is as old as humanity itself. Every occupation or conquest on a coast may be said to carry with it the presumption to a further inland claim. The Power that occupied Cairo, or built Calcutta, was thereby committed to an advance that could not stop at the deltaic region. The famous controversies between the United States and Spain as to the boundaries of Louisiana, after the cession of the latter to America by France in 1803, and between the United States and Great Britain over the Oregon Territory, revolved round the question of the rights conferred by discovery or settlement. At the Berlin Congress, Bosnia and Herzegovina, though inhabited by a Slav people, were handed over to Austria because she already possessed Dalmatia.

But it is, again, for the most part, in Africa, arising out of the emulous descent upon its coasts of the principal European Powers, that the doctrine of Hinterland in its modern aspect has taken formal shape. A double question arises in the case of such occupations, namely, how far they may be supposed to extend laterally, and how far inland. The latter is the Hinterland problem. It is often held that the inland Frontier should extend to the water-divide of the rivers debouching within the line of coast occupation. But any such principle must be open to many exceptions: and the actual extent of Hinterland that is held to belong to any Power depends, in the main, upon the degree to which it succeeds in rendering its authority effective in the interval before it is fixed by international agreement. A forward step in the regularization of coast occupation in Africa was taken by the Agreement of the leading

Powers in the Berlin Conference of 1885, requiring the notification of any such action in the future to the Signatory Powers, in order to enable them to substantiate any counterclaim of their own, and stipulating for the effective exercise of authority in the region concerned. This Agreement only applied to Africa, nor even there did it cover the interior extension of Frontiers. But it has not been without influence in imparting some measure of propriety to proceedings not everywhere over-imbued with scruple. The most recent instances in which the Hinterland doctrine has been before the public have been the dispute between Great Britain and Venezuela as to the inland boundary; the provisions by which the Great Powers, when leasing naval bases on the coast of China, acquired at the same time an interior zone; and the steps taken a short while ago to define, by means of an Anglo-Turkish Boundary Commission, the Hinterland of Aden, where Turkish troops from the Yemen were constantly encroaching upon the tribes within the British Protectorate.

The reference to China invites attention to yet another form of Frontier extension that has found favour in recent times. This is the grant of Leases, in order to veil an occupation not as a rule intended to be temporary. Great Britain has sometimes made use of this expedient in Native States in India, Quetta, and afterwards Nushki, having been taken from the Khan of Kelat on a quit rent in perpetuity. I also, while Viceroy, negotiated the perpetual Lease of the interior province of Berar by the Nizam of Hyderabad, though this was an act exclusively of administrative and financial convenience, and had nothing to do with exterior Frontiers. Some of the territories of the Sultan of

Zanzibar were leased, first for a term of years and afterwards in perpetuity, to Great Britain and Germany. Port Arthur, Wei-Hai-Wei, Kiao Chau, and Kowloon are cases in which the fiction of a Lease has been employed to cover what might otherwise appear to have been a violation of the territorial integrity of China. Sometimes Leases are granted for commercial purposes, or as part of a diplomatic bargain. Such has been the case with the so-called Enclaves, leased by Great Britain to King Leopold on the Nile and to France on the Niger. Experience, though not as yet very old, shows that the tendency of Leases is, from being temporary to become permanent, and, in fact, to constitute a rudimentary form of ulterior possession.

Of these modern expedients the last to be noticed represents the most shadowy form of the Sphere of Influence that has yet been devised by the ingenuity of modern diplomacy. I refer to the promise made by a weaker Power to a stronger not to alienate by lease, mortgage, or cession a specified portion of its territories to any other Power. This does not necessarily, though it may sometimes, imply the exercise of a Protectorate by the stronger of the two contracting parties,[1] but it tacitly recognizes some sort of reversionary claim on the part of the latter. At the weakest, it is a sort of diplomatic manifesto to other Powers of a special degree of interest entertained by one. Great Britain's desire to earmark as a potential Sphere of Influence the valley

---

[1] Undoubtedly in the Persian Gulf the conclusion of such Agreements by the Indian Government with the Trucial Chiefs and with the Sheikh of Koweit, was tantamount to an assertion of Protectorate, although in the latter case, by a strange anomaly, the Protectorate of Turkey was never formally denied.

of the Yangtsze in China did not proceed beyond this somewhat impalpable assertion, which was promptly challenged by Germany. A peculiar feature of these arrangements is that the ruler or State who gives the self-denying pledge very often does so under the minimum of pressure, and sometimes with ill-concealed delight. The perils or chances of future deprivation appear to be remote : in the interim his own title to ownership has been reaffirmed by a great Power, and in this fact a useful protection may be sought against the designs or encroachments of other interested parties.

Of all the diplomatic forms or fictions which have latterly been described, it may be observed that the uniform tendency is for the weaker to crystallize into the harder shape. Spheres of Interest tend to become Spheres of Influence ; temporary Leases to become perpetual ; Spheres of Influence to develop into Protectorates ; Protectorates to be the forerunners of complete incorporation. The process is not so immoral as it might at first sight appear ; it is in reality an endeavour, sanctioned by general usage, to introduce formality and decorum into proceedings which, unless thus regulated and diffused, might endanger the peace of nations or too violently shock the conscience of the world. I know of no more striking illustration of this tendency than the development of Lord Salisbury's Siamese Declaration of January, 1896, by which the single and uncontested authority of Siam over the unguaranteed Siamese territory lying outside of the Menam watershed was specifically affirmed, with Lord Lansdowne's Declaration of April, 1904, by which this territory was openly divided into French and British Spheres of Influence, in which the two Powers mutually conceded to each other liberty of action. Lord Salisbury's was the first step : this was the second : and if at any time there is a third, its approximate character can be foreseen.

Evidences of Progress.

The recent portions of my lecture may have suggested the suspicion that modern nations, in the extension of their Frontiers, are not only not more scrupulous, but are more crafty in their methods than rulers of former times. Such would, I think, be a mistaken impression. About the respective ethical pretensions of the ancient and modern world I will not here dispute. But I regard it as incontestable, not merely that the expedients which I have described make on the whole for peace rather than for war—which is perhaps a sufficient vindication of them : but that there has been a progressive advance, both in the principles and in the practice of Frontier policy, which has already exercised a widespread effect, and is of good omen for the future. Let me, before I conclude, collect the evidences of this amelioration.

The history which I have narrated shows the immense increase in the number and diversity of the Frontiers that have been adopted to protect the possessions and to control the ambitions of States. The primitive forms, except where resting upon indestructible natural features, have nearly everywhere been replaced by boundaries, the more scientific character of which, particularly where it rests upon treaty stipulations, and is sanctified by International Law, is undoubtedly a preventive of misunderstanding, a check to territorial cupidity, and an agency of peace.

But there has been a much greater and more beneficent advance in the machinery and implements employed than in the nature or diversity of the Frontiers chosen. In the first place the idea of a demarcated Frontier is itself an essentially modern conception, and finds little or no place in the ancient world. In Asia, the oldest inhabited continent, there has always been a strong instinctive aversion to the acceptance of fixed boundaries, arising partly from the nomadic habits of the people, partly from the dislike of precise arrangements that is typical of the oriental mind, but more still from the idea that in the vicissitudes of fortune more is to be expected from an unsettled than from a settled Frontier. An eloquent commentary on these propensities is furnished by the present position of the Turco-Persian Frontier, which was provided for by the mediation of Great Britain and Russia in the Treaty of Erzerum exactly sixty years ago, and was even defined, after local surveys, by Commissioners of the two Powers as existing somewhere in a belt of land from 20 to 40 miles in width stretching from Mount Ararat to the Persian Gulf. There, unmaterialized and unknown, it has lurked ever since, both Persia and still more Turkey finding in these unsettled conditions an opportunity for improving their position at the expense of their rival that was too good to be surrendered or curtailed. In Asiatic countries it would be true to say that demarcation has never taken place except under European pressure and by the intervention of European agents.

But even in Europe, where fixed boundaries are of much older standing, it is surprising to note the absence or inadequacy till recent times of proper arrangements for calling them into being. The earliest instance of a Frontier Commission that I have come across is that of the Commission of six English and Scotch representatives who were appointed in 1222 to mark the limits of the two kingdoms, and it is symptomatic of the contemporary attitude about Frontiers that it broke down directly it set to work, leaving behind it what became a Debatable Land and a battle-ground of deadly strife for centuries. Even in the seventeenth-century treaties, by which the map of Europe was practically reconstructed, there is no express provision for demarcation. It is not till after the middle of the eighteenth century that we find Commissioners alluded to in the text of treaties, and reference made to topographical inquiries and surveys of engineers. What seems to us now the first condition of a stable Frontier appears to have been then regarded as the least important. Perhaps one reason was that no one expected and few desired that stability should be predicated of any political Frontier.

Contrast this with the methods now employed. Local surveys or reconnaissances, where one or the other has been found possible, precede the discussions of statesmen. Small Committees of officials are frequently appointed in advance to consider the geographical, topographical, and ethnological evidence that is forthcoming, and to construct a tentative line for their respective Governments; this, after much debate, is embodied in a treaty, which provides for the appointment of Commissioners to demarcate the line upon the spot and submit it for ratification by the principals. Geographical knowledge thus precedes or is made the foundation of the labours of statesmen, instead of supervening at a later date to cover them with ridicule or reduce their findings to a nullity. I do not say that absurd mistakes and blunders are not still committed. I could, if I had the leisure, construct a notable and melancholy list. But the tendency is unquestionably in the direction of greater precision both of knowledge and of language.

Lastly, when the Commissioners reach the locality of demarcation, a reasonable latitude is commonly conceded to them in carrying out their responsible task. Provision is made for necessary departures from the Treaty line, usually 'on the basis of mutual concession'; tribes or villages are allowed to use watering places or grazing grounds across the Frontier, or to choose on which side of the border they will elect to dwell. Some Treaties (for instance that between the United States and Mexico) allow for the pursuit of raiders across the common Frontier without the creation of a *casus belli*. When the Commissioners have discharged their duty, not as a rule without heated moments, but amid a flow of copious hospitality and much champagne, beacons or pillars or posts are set up along the Frontier, duly numbered and recorded on a map. The process of demarcation[1] has in fact become one of expert labour and painstaking exactitude.

---

[1] I use the word intentionally as applying to the final stage and the marking out of the boundary on the spot. Diplomatic agents and documents habitually confound the meaning of the two words 'delimitation' and 'demarcation', using them as if they were interchangeable terms. This is not the case. Delimitation signifies all the earlier processes for determining a boundary, down to and including its embodiment in a Treaty or Convention. But when the local Commissioners get to work, it is not delimitation but demarcation on which they are engaged.

A further development has taken place in the personality and qualifications of Commissioners. Commonly these are carefully selected representatives of the two Powers directly concerned. Occasionally in Asia, and almost invariably in Africa, the curious phenomenon is witnessed, sometimes under Treaty stipulations, as a rule independently of them, of the demarcation of boundaries by Commissioners drawn not from the country directly affected but from the great Powers between or within whose spheres of influence it may lie. Thus Great Britain and Russia determined on their own account the north-west Frontier of Afghanistan in 1886. Where native agents are admitted, usually in a subordinate and advisory capacity, they are apt to interpret their functions as justifying an exceptional measure of vacillation, obstruction, and every form of delay. Any one who has had experience of demarcation on the Frontiers of Persia and Afghanistan will recall the prodigies that are capable of being performed in these directions. Sometimes, after an International Agreement, such as the Treaty of Berlin, the Frontiers there laid down are demarcated by European Commissioners, officers of the highest technical knowledge and repute being nominated by the several Powers.

All of these marks of progress however, practical and valuable as they are, shrink into relative insignificance when compared with the startling change introduced by the reference of Frontier questions to arbitration by external individuals or States. This method is the exclusive creation of the last half-century or less, and its scope and potentialities are as yet in embryo. Sometimes the reference may be to a Sovereign or jurist of international reputation: the rulers of the more detached or less powerful States, such as the Kings of Sweden and Italy or the President of the Swiss Confederation, being much in request; in one case, that of Venezuela, the United States were admitted as *amicus curiae*, acting on behalf of the smaller State of which they had assumed the virtual protectorate. Treaties providing for arbitration in a number of cases which may embrace Frontiers have been concluded between several of the leading Powers.

Finally, in the last decade, there has been set up the International Tribunal of the Hague, which, if its prestige be maintained and a permanent Court established, will probably become in an increasing degree the referee and arbiter of the Frontier disputes of the future.

These symptoms, when viewed in combination, will, I think, be sufficient to justify the claim that progress in the delimitation of Frontiers is positive and real. It would be futile to assert that an exact Science of Frontiers has been or is ever likely to be evolved: for no one law can possibly apply to all nations or peoples, to all Governments, all territories, or all climates. The evolution of Frontiers is perhaps an art rather than a science, so plastic and malleable are its forms and manifestations. But the general tendency is forward, not backward; neither arrogance nor ignorance is any longer supreme; precedence is given to scientific knowledge; ethnological and topographical considerations are fairly weighed; jurisprudence plays an increasing part; the conscience of nations is more and more involved. Thus Frontiers, which have so frequently and recently been the cause of war, are capable of being converted into the instruments and evidences of peace.

There are many other branches of the subject upon which I should like to have dwelt to-day, but which I have not here the time to examine. Such are the reciprocal influence of Fortifications upon Frontiers, and of Frontiers upon Fortifications; the effect upon Frontiers of modern scientific inventions, such as the electric telegraph, railroads and tunnels, and munitions of war; the experience and romance of Frontier Commissions. There is also a class of so-called Natural Frontiers which I have been obliged to omit, as possessing no valid claim to the title, namely those which are claimed by nations as natural on grounds of ambition, or expediency, or more often sentiment. The attempt to realize Frontiers of this type has been responsible for many of the wars, and some of the most tragical vicissitudes in history. But its treatment would almost demand an independent Essay. When I began to write this lecture I had further contemplated tracing the comparative evolution of the Frontiers of all the great Empires in history, giving an exact account of the Indian Frontier system, at present the most highly organized in the world, and comparing it point by point with its ancient counterpart and prototype, the Frontier system of Rome. These designs must be postponed for a larger work on the same subject, if the leisure for this be ever found. Another deferred topic is the engrossing subject of Border Literature, in which it would, I believe, be possible to demonstrate a common growth and characterization in diverse periods and many lands. I should also like to have analysed the types of manhood thrown up by Frontier life, savage, chivalrous, desperate, adventurous, alluring. To only one of these allied subjects will I refer before I conclude, and that is the influence of Frontier expansion upon national character, as illustrated in the history of the Anglo-Saxon race.

We may observe two very distinct types of this influence on the eastern and western sides of the Atlantic. A modern school of historians in America has devoted itself with patriotic ardour to tracing the evolution of the national character as determined by its western march across the continent. In no land and upon no people are the evidences more plainly stamped. Not till the mountains were left behind and the American pioneers began to push across the trackless plains, did America cease to be English and become American. In the forests and on the trails of the Frontier, amid the savagery of conflict, the labour of reclamation, and the ardours of the chase, the American nation was born. There that wonderful and virile democracy, imbued with the courage and tenacity of its forefathers, but fired with an eager and passionate exaltation, sprang into being. The panorama of characters and incidents, already becoming ancient history, passes in vivid procession before our eyes. First comes the trapper and the fur trader tracking his way into the Indian hunting-grounds and the virgin sanctuaries of animal life. Then the backwoodsman clears away the forests and plants his log hut in the clearings. There follow him in swift succession the rancher with his live-stock, the miner with his pick, the farmer with plough and seeds, and finally the urban dweller, the manufacturer, and the artisan. On the top of the advancing wave floats a scum of rascality that is ultimately deposited in the mining camps of California and the gambling dens of the Pacific Coast. Scenes of violence and carnage, the noise of fire-arms, and the bleaching bones of men, mark the advance. The voice of the backwoods-preacher sounds through the tumult in accents of mingled ecstasy and rebuke. But from this tempestuous cauldron of human passion and privation, a new character, earnest, restless, exuberant, self-confident, emerged; here an Andrew Jackson, there an Abraham Lincoln, flamed across the stage; and into this noble heritage of achievement and suffering, the entire nation, purified and united in its search for the Frontier, both of its occupation and its manhood, has proudly entered.

Now let us turn to the other side of the world, where on a widely different arena, but amid kindred travail, the British Empire may be seen shaping the British character, while the British character is still building the British Empire. There, too, on the manifold Frontiers of dominions, now amid the gaunt highlands of the Indian border, or the eternal snows of the Himalayas, now on the parched sands of Persia or Arabia, now in the equatorial swamps and forests of Africa, in an incessant struggle with nature and man, has been found a corresponding discipline for the men of our stock. Outside of the English Universities no school of character exists to compare with the Frontier; and character is there moulded, not by attrition with fellow men in the arts or studies of peace, but in the furnace of responsibility and on the anvil of self-reliance. Along many a thousand miles of remote border are to be found our twentieth-century Marcher Lords. The breath of the Frontier has entered into their nostrils and infused their being. Courage and conciliation—for unless they have an instinctive gift of sympathy with the native tribes, they will hardly succeed—patience and tact, initiative and self-restraint, these are the complex qualifications of the modern school of pioneers. To these attainments should be added—for the ideal Frontier officer—a taste for languages, some scientific training, and a powerful physique. The work, which he may be called upon to perform, may be that of the explorer or the administrator or the military commander, or all of them at the same time. The soldier, perhaps more often than the civilian, furnishes this type; and it is on the Frontier that many of the greatest military reputations have been made. The Frontier officer takes his life in his hands; for there may await him either the knife of the Pathan fanatic, or the more deadly fevers of the African swamp. But the risk is the last thing of which he takes account. He feels that the honour of his country is in his hands. I am one of those who hold that in this larger atmosphere, on the outskirts of Empire, where the machine is relatively impotent and the individual is strong, is to be found an ennobling and invigorating stimulus for our youth, saving them alike from the corroding ease and the morbid excitements of Western civilization. To our ancient Universities, revivified and reinspired, I look to play their part in this national service. Still from the cloistered alleys and the hallowed groves of Oxford, true to her old traditions, but widened in her activities and scope, let there come forth the invincible spirit and the unexhausted moral fibre of our race. Let the advance guard of Empire march forth, strong in the faith of their ancestors, imbued with a sober virtue, and above all, on fire with a definite purpose. The Empire calls, as loudly as it ever did, for serious instruments of serious work. The Frontiers of Empire continue to beckon. May this venerable and glorious institution, the nursery of character and the home of loyal deeds, never fail in honouring that august summons.

OXFORD
PRINTED AT THE CLARENDON PRESS
BY HORACE HART, M A.
PRINTER TO THE UNIVERSITY

Printed in Great Britain by
Amazon.co.uk, Ltd.,
Marston Gate.